Learning from TIMSS

Results of the Third International Mathematics and Science Study

Summary of a Symposium

Alexandra Beatty, *Editor*

Board on International Comparative Studies in Education
Board on Testing and Assessment
Commission on Behavioral and Social Sciences and Education

Committee on Science Education K-12
Mathematical Sciences Education Board
Center for Science, Mathematics, and Engineering Education

National Research Council

NATIONAL ACADEMY PRESS
Washington, D.C. 1997

NOTICE: The project that is the subject of this report was approved by the Governing Board of the National Research Council, whose members are drawn from the councils of the National Academy of Sciences, the National Academy of Engineering, and the Institute of Medicine. The members of the committee responsible for the report were chosen for their special competences and with regard for appropriate balance.

This report has been reviewed by a group other than the authors according to procedures approved by a Report Review Committee consisting of members of the National Academy of Sciences, the National Academy of Engineering, and the Institute of Medicine.

The National Academy of Sciences is a private, nonprofit, self-perpetuating society of distinguished scholars engaged in scientific and engineering research, dedicated to the furtherance of science and technology and to their use for the general welfare. Upon the authority of the charter granted to it by the Congress in 1863, the Academy has a mandate that requires it to advise the federal government on scientific and technical matters. Dr. Bruce M. Alberts is president of the National Academy of Sciences.

The National Academy of Engineering was established in 1964, under the charter of the National Academy of Sciences, as a parallel organization of outstanding engineers. It is autonomous in its administration and in the selection of its members, sharing with the National Academy of Sciences the responsibility for advising the federal government. The National Academy of Engineering also sponsors engineering programs aimed at meeting national needs, encourages education and research, and recognizes the superior achievements of engineers. Dr. William A. Wulf is president of the National Academy of Engineering.

The Institute of Medicine was established in 1970 by the National Academy of Sciences to secure the services of eminent members of appropriate professions in the examination of policy matters pertaining to the health of the public. The Institute acts under the responsibility given to the National Academy of Sciences by its congressional charter to be an adviser to the federal government and, upon its own initiative, to identify issues of medical care, research, and education. Dr. Kenneth I. Shine is president of the Institute of Medicine.

The National Research Council was organized by the National Academy of Sciences in 1916 to associate the broad community of science and technology with the Academy's purposes of furthering knowledge and advising the federal government. Functioning in accordance with general policies determined by the Academy, the Council has become the principal operating agency of both the National Academy of Sciences and the National Academy of Engineering in providing services to the government, the public, and the scientific and engineering communities. The Council is administered jointly by both Academies and the Institute of Medicine. Dr. Bruce M. Alberts and Dr. William A. Wulf are chairman and vice chairman, respectively, of the National Research Council.

This report was supported by Cooperative Grant No. OSR-9355774 from the National Science Foundation, which includes funds from the National Center for Education Statistics of the U.S. Department of Education. Any opinions, findings, conclusions, or recommendations expressed in this publication are those of the author(s) and do not necessarily reflect the views of the organizations or agencies that provided support for this project.

International Standard Book Number 0-309-05975-5

Additional copies of this report are available from:
National Academy Press
2101 Constitution Avenue, N.W.
Lock Box 285
Washington, D.C. 20055
Call 800-624-6242 or 202-334-3313 (in the Washington Metropolitan Area).

This report is also available on line at **http://www.nap.edu**

Printed in the United States of America

BOARD ON INTERNATIONAL COMPARATIVE STUDIES IN EDUCATION

BOARD ON TESTING AND ASSESSMENT

COMMITTEE ON SCIENCE EDUCATION K-12

Jane Butler Kahle (*Chair*), Department of Teacher Education, Miami University
Joseph McInerney (*Vice Chair*), Biological Sciences Curriculum Study, Colorado Springs, Colorado
J. Myron Atkin, Center for Educational Research, Stanford University
Caryl Edward Buchwald, Department of Geology, Carleton College
George Bugliarello, Chancellor, Polytechnic University
Christine Chopyak-Minor, Keystone Science School, Keystone, Colorado
Peter B. Dow, Buffalo Museum of Science, New York
William E. Dugger, Jr., Technology for All Americans, Blacksburg, Virginia
Wade Ellis, Jr., Department of Mathematics, West Valley College, Saratoga, California
Norman Hackerman, The Robert A. Welch Foundation, Houston, Texas
William D. Hammers, Optimal Solutions, Clearwater, Kansas
Robert Hazen, Carnegie Institution of Washington, Washington, D.C.
Leroy Hood, Department of Molecular Biotechnology, University of Washington, Seattle
Michael G. Lang, Phoenix Urban Systemic Initiative, Maricopa Community College
William Linder-Scholer, SciMathMN, St. Paul, Minnesota
Maria Alicia Lopez-Freeman, Center for Teacher Leadership in Language and Status, California Science Project, Monterey Park
John A. Moore, Department of Biology, University of California, Riverside
Darlene Norfleet, Flynn Park Elementary School, University City, Missouri
William Spooner, North Carolina Department of Public Instruction, Raleigh
Judith Sydner-Gordon, Los Angeles County Office of Education, California
Rachel Wood, Delaware State Department of Public Instruction, Dover

Harold Pratt, *Director*

MATHEMATICAL SCIENCES EDUCATION BOARD

Contents

Learning from TIMSS

Results of the Third International Mathematics and Science Study

INTRODUCTION

Economic and technological links among nations have mushroomed during the past two decades and are manifest in entities as large as multinational corporations and as modest as Internet chatrooms. Such links reveal the existence of—and foster curiosity about—differences and similarities among nations, particularly with regard to endeavors common to all nations, such as education. A number of international comparative studies conducted in recent years, as well as other evidence, have shown that education systems vary substantially. A careful look at other systems can both deepen any country's understanding of its own educational beliefs and methods and introduce new possibilities. Researchers, policy makers, teachers, parents, and others would like answers to a variety of questions. What do other countries do, and how do they do it? How effective are they in improving achievement? In what ways is the U.S. system like others? How is it different? How might it be strengthened? What does the United States do that other nations want to emulate?

By far the most ambitious exploration of questions such as these to date is the Third International Mathematics and Science Study (TIMSS), which was conducted under the auspices of the International Association for the Evaluation of Educational Achievement (IEA). Nearly 50 countries participated in various parts of TIMSS; materials were developed in more than 30 languages for use in the study. More than half a million students at three age levels (9, 13, 17) from 15,000 schools participated in the study, and students, teachers, and administrators in more than 20 countries responded to background questionnaires designed to elicit contextual information. Several auxiliary studies were also conducted. Researchers evaluated and compared the curricula of nearly 50 countries; experts observed and analyzed a subset of school systems in Japan, Germany, and the United States; and a videotape study of classroom lessons in the same three countries was conducted.

Planning for the study began in 1991 and data were collected in 1995 and 1996. The first set of primary analyses, covering 13-year-olds in 41 countries, was released in late 1996; the analyses for the 9-year-olds were released in mid-1997; the last set of primary analyses, for 17-year-olds, is to be released in 1998.

TIMSS has yielded an unprecedented body of data with which to explore both targeted questions about mathematics and science achievement and larger questions about the structure and curricular goals of education systems in different nations. However, the very magnitude of the study, the newness of some of the research methodology, and persistent pressure to translate complex information into simple conclusions all raise concerns about the research methodology and about the implications of the study findings for policy decisions.

To begin to address these questions, and to encourage innovative and far-sighted exploration of TIMSS resources, the National Research Council held a symposium in Washington, D.C., on February

3-4, 1997. The primary goal of the symposium was to seize a moment—soon after the initial release of findings from TIMSS—when many of the central concerns regarding mathematics and science education would be the focus of considerable public attention. The symposium was designed to "complicate" a discussion that could easily be oversimplified: to foster appreciation of the study's complexity and of the range and depth of analyses it makes possible. Assuming that the "horse-race" rankings of nations made possible by the achievement results would receive the greatest publicity when the data were released, the symposium planners wanted to initiate a sustained discussion of the data, as well as encourage collaboration among communities of scholars. By raising awareness of some of the difficult issues presented by the complexity of the study's design, they hoped to influence the ongoing discussion of the study in ways that would enhance its potential to advance education reform. Recognizing the magnitude of the study itself and the multitude of issues it raises, they intended to encourage others to continue this discussion, not to complete it in one session.

Participants included officials from the U.S. Department of Education, representatives from many private institutions concerned with education issues broadly and with mathematics and science education in particular, investigators who have been involved with TIMSS, researchers, and representatives from various professional groups. (See the list of participants in Appendix A.) The symposium was sponsored by four boards of the National Research Council: the Board on International Comparative Studies in Education, the Board on Testing and Assessment, the Committee on Science Education K-12, and the Mathematical Sciences Education Board. Support came from the U.S. National Science Foundation and the National Center for Education Statistics (NCES) of the U.S. Department of Education.

The symposium had two major components: a detailed look at TIMSS itself and the beginnings of a critical discussion of issues raised by it. The principal researchers responsible for the four major components of the study described their work and highlighted a few of their key findings and some of the methodological challenges they faced. Discussants for each of these sessions, as well as participants, raised issues of interpretation, use, and application of the study data. The remaining sessions were designed to look critically at several aspects of the study and to provide a variety of perspectives on the study and the role it might play in policy planning. Although some of the presenters addressed critiques of aspects of the study, the symposium was not designed to provide a thorough critical analysis of TIMSS; rather, it was designed to focus on issues relevant to TIMSS's implications for the future. (See the symposium agenda in Appendix B.)

Five scholars prepared papers for the symposium. Each was asked to reflect critically on either a particular aspect of the study itself or some of its implications in the current policy environment. The resulting presentations and discussions ranged widely—from close scrutiny of methodological questions to intense consideration of the structure of public education in the United States. As the symposium planners intended, the presentations and discussion focused not on achieving group consensus, but on unearthing a variety of views on a complex topic. (See Appendix C for a list of the papers presented.) Clearly the day and a half allotted for the symposium did not allow for an exhaustive discussion of either the strengths and weaknesses of TIMSS or its many implications for policy makers. Moreover, because of time constraints, a number of important points were raised but not elaborated during the discussion.

This summary report is an additional component of the effort to foster dialogue in the education research and policy communities. It describes the major elements of TIMSS, presents some of the discussion that took place at the symposium, and explores the themes that emerged from it. Because TIMSS is so complex, the steering committee charged with planning the symposium decided to devote considerable symposium time to explication of the structure of the study and a few of its principal findings. This document follows that lead. The next section, "What Is TIMSS?," provides a description of the study and of the presentations made by the TIMSS researchers. The following two sections summarize, respectively, the questions and critiques that presenters raised about the study itself and the major policy issues that were addressed. The last section summarizes the major ideas that emerged at the symposium.

WHAT IS TIMSS?

As its name indicates, TIMSS is the third in a series of investigations of mathematics and science learning conducted under the auspices of the International Association for the Evaluation of Educational Achievement. IEA is an international consortium of research institutions in more than 40 countries. Although individual governments may fund their countries' participation in IEA activities, the organization is run by an assembly of country representatives. The first IEA study, of mathematics, was conducted in the 1960s; the second mathematics study was done in the 1970s. IEA has also conducted studies of learning in a variety of other subjects. Although the structure and composition of IEA's studies have evolved some since the 1960s, their purpose—to describe and explain differences in student achievement—has remained the same.

More specifically, the organizers of the study described the purpose of TIMSS in this way: "to learn more about mathematics and science curricula and teaching practices associated with high levels of student achievement, in order to improve the teaching and the learning of mathematics around the world" (Robitaille and Garden, 1996:15). Study planners recognized that to accomplish this goal they would need to collect a variety of different kinds of data. First, they needed the kind of common measure of achievement used in previous studies—numbers that would represent the varying degrees to which students around the world have learned the body of mathematics and science knowledge deemed (through international consensus) essential. This was obtained by means of an achievement test (described in greater detail below). All of the other components of TIMSS were designed to provide data that can help explain variations in performance on the achievement test: these included a detailed look at the content of mathematics and science curricula and textbooks around the world, as well as investigations of student attitudes and experiences, teaching practices and school resources, and many other factors that affect achievement (these other components of the study are described below). The challenge for TIMSS researchers, and for others wishing to use the data for additional analyses, is to make full use of this combination of information about the education practices and contexts that influence student learning.

The scope of TIMSS is unprecedented in several ways. Though many international comparative assessments have been conducted, none has assessed student learning in two subjects in so many countries at the same time. Those involved in the planning and design of the study paid considerable attention to the experience gained in the study's predecessor, the Second International Mathematics Study (SIMS) (McKnight et al., 1987; Medrich and Griffith, 1992). They addressed many of the criticisms leveled at SIMS, both by adhering to strict sampling procedures and by expanding the scope of the design for TIMSS to include the collection of an extensive variety of contextual data (Rotberg, 1990; Bracey, 1996; Third International Mathematics and Science Study, 1996). In addition, the designers of TIMSS incorporated research methods from several different disciplines in a groundbreaking effort to link different kinds of data. Essentially, several distinct studies were conducted, each investigating questions about mathematics and science learning from a different perspective. The combination of different research methods raised a variety of issues and questions, some of which are addressed below ("Critiques and Mathodological Issues"). (See Appendix D for a bibliography of TIMSS reports and resources.)

The different components of the study grew out of three basic questions that it was designed to answer: What are students in each nation expected to learn? What, and how, are students actually taught? What do students actually learn? TIMSS researchers used the terms "intended, implemented, and achieved curricula," respectively, to refer to these three basic questions (Robitaille and Garden, 1996).

The Achievement Study

The core of TIMSS is an assessment of student achievement in mathematics and science, administered to students at ages 9 (Population 1), 13 (Population 2), and 17 (Population 3). The achievement results, of course, provide the data on the achieved curriculum—what students have actually learned. The content to be tested in each subject and at each age level was determined through a sometimes contentious consensus process involving all of the participating countries. The resulting framework document, which guided the development of the test questions, reflects many compromises; it does not reflect the actual curriculum in any one country, and each country is free to conduct further analyses on just those questions that covered the material taught to its own students (International Association for the Evaluation of Educational Achievement, 1996a, 1996b).

The test itself is similar to other large-scale assessments that are used in the United States, such as the National Assessment of Educational Progress (NAEP). It is a combination of multiple-choice questions and open-ended exercises that ask students to generate solutions to problems or to answer questions in their own words. The open-ended exercises are scored using guidelines that describe several categories of responses and assign scores to them. In each country the test was administered to a sample of classes of students—approximately 3,750 students per country at each grade level (Third International Mathematics and Science Study, 1996). The samples were chosen so that various groups were adequately represented and each country's overall population characteristics were reflected. Each student answered only a portion of the questions meant for his or her grade level; various subsets of the questions were printed in different test booklets so that an appropriate number of students in each sample would take each possible combination of questions. Consequently, data could be reported on the entire content domain covered by the test although each student sat for only 60 or 90 minutes of testing. The complex item sampling design made it possible for researchers to report on the performance of different population groups and on student performance for different types of questions and different content areas. The sampling procedure also made possible the so-called "horse race" results, which rank the performances of participating countries. Results are being reported for nations and, in the United States, for three states and one consortium of school districts.[1] Forty-one countries participated in the assessment of middle-school, or Population 2, students (13-year-olds); these results were released shortly before the symposium. Twenty-six nations participated in the elementary school, or Population 1, portion (9-year-olds), results for which were released in June 1997. Data for Population 3, students at the end of secondary school (17-year-olds), are scheduled for release in February 1998.[2] No individual scores are available.

[1]The three states, Colorado, Illinois, and Minnesota, and the First in the World Schools, a consortium in the northwest suburbs of Chicago, provided funds for their participation as "mini-nations" in order to learn how their own students compare to others internationally. NCES has made it possible for other states or districts who wish to administer TIMSS locally to do so.

[2]Symposium participants repeatedly stressed the importance of recognizing, when drawing interpretations from TIMSS, that different groups of nations participated in different portions of the project. See note 5 on page 17 for the numbers of countries participating in each major component.

Because the results are based on the performance of representative samples of students in each country, they actually, as TIMSS researchers explained, "represent a range within which the nation's actual average would most likely fall if all students were tested" (TIMMS U.S. National Research Center, 1996). Thus, the U.S. achievement results were presented in terms of three bands—groups of countries that performed better than the United States did, at approximately the same level as the United States, or worse than the United States. By presenting the results this way, researchers hoped to discourage observers from focusing on slight differences that might be inappropriately magnified if numerical scores were simply listed in rank order.

More than 20 countries also chose to include a set of performance assessment tasks for Populations 1 and 2; these were simple experiments using standardized materials provided in kits. The tasks were too expensive and time-consuming to include for the entire testing population, but they are expected to yield data on skills not easily measured by paper-and-pencil assessments (National Center for Education Statistics, 1996). Testing of Population 3 students also addressed two "specialist" subpopulations: students enrolled in advanced mathematics or physics courses.

Background Questionnaires

At the time the assessments were administered, students, teachers, and school officials were also asked to fill out background questionnaires designed to elicit important information about the contexts in which student learning occurs. These questionnaires collected data on students' and teachers' backgrounds, school structures and resources, students' and teachers' attitudes about mathematics and science, teachers' pedagogical beliefs and practices, classroom coverage of various mathematics and science topics, and other variables. Responses to these questions can then be correlated with achievement data to reveal associations between various factors and student performance. Although such associations cannot support specific causal inferences, they can call attention to factors that are associated with success and identify promising areas for further study.

Quality Control

The planners for TIMSS took great care to ensure the quality of the data collection, and independent observer Edward Haertel commented on the high quality of the sampling and data collection in the paper he presented at the symposium. The research team paid particular attention to the sampling in part because SIMS, its predecessor, was criticized for using sampling methods that may have distorted the international comparisons. An entire volume documenting the quality control procedures used in TIMSS has been published (Third International Mathematics and Science Study, 1996), but it is worth noting one strategy in particular. Because the sampling rules were so rigorous and complicated, not all countries were able to meet all of them, but the data collected from these countries were still of value. The TIMSS research team defined several levels of compliance, which were clearly indicated in the main ranking tables. Thus, readers could see easily that comparisons between nations with differing levels of compliance should be made with caution and with an understanding of the nature of these differing levels.

Albert Beaton, TIMSS study director, presented a brief summary of the study at the symposium and highlighted a few of the key results from the Population 2 data, the first data to be released and the only data available at the time of the symposium.[3] He began by noting that, while there has been worldwide interest in the country rankings, members of the press had not really addressed the more complex findings of the study or the issues and the questions they raise. For example, Beaton showed a table depicting results for the 41 Population 2 countries, similar to those used in the published reports. He explained that a reporter from a national news magazine had declined to publish it on the grounds that it was too complicated. Perhaps the most striking finding for Beaton was that all of the reporting countries show a connection between socioeconomic factors and performance. In every one of the 41 countries, he explained, "there is a relationship between the number of books in the home and school performance." There was a similarly clear relationship across countries tested between parents' levels of education and student performance. Other factors explored in TIMSS did not demonstrate such clear relationships: for example, class size shows some relationship to achievement, except that Korea, whose performance was second only to Singapore's, averages more than 40 students per class.

Beaton presented some other key findings:

- There are differences in performance on particular content areas covered by the assessment that are consistent with differences in curricula across countries.
- U.S. seventh-grade students ranked higher among nations than did U.S. eighth-grade students. Beaton remarked that this finding is important because it supports the overall achievement differences that were found. That is, differences between grades within a nation cannot be explained away by a large national difference, which would have affected performance at both grades equally.
- Within most countries and overall, boys had significantly higher mean science achievement than did girls in both the seventh and eighth grades. Gender differences in mathematics achievement were small or nonexistent; differences that did exist favored boys.
- There is a large difference in average science and mathematics achievement between the top-performing and bottom-performing

[3]Population 2 covered the two school grades containing the largest numbers of 13-year-olds, grades seven and eight in the United States.

countries. Despite this large difference, when countries were ordered by average achievement, there were only small differences in achievement between each country and the ones closest to it.

- In science, students generally had the most difficulty with the chemistry items.
- In mathematics, the questions that stood out as most difficult called for multistep problem solving and applications.
- In both mathematics and science, country performance in different content areas seemed to correspond to curricular emphasis.

Beaton was the first of many at the symposium to point out that the TIMSS data has, not surprisingly, failed to produce a "silver bullet" that will magically transform mathematics and science education. As Beaton put it: "Wouldn't it be nice to just find that all we have to do is something simple, you know, increase the school year, for example? . . . We have been poring over the data . . . and there is just no simple answer." For every likely looking connection between achievement and a variable such as amount of homework or class size, TIMSS showed counterexamples. Beaton and his colleagues concluded that, while each probably has an effect, none by itself made a major difference.

The Curriculum Study

As even casual observation reveals, there are substantial differences among the education systems and curricula in use in the participating nations. The purpose of the curriculum study was to find a way to make sense of these differences and to make it possible to explore the relationship between curriculum and achievement results. More specifically, researchers hoped that by looking systematically at which topics are covered at which levels around the world, and at performance expectations, they could gain understanding of differences in student performance on particular skills and segments of the content that were tested. This study, of course, primarily explored what study planners called the intended curriculum.

Undertaking a thorough comparison among the curricula of 46 countries was complicated by the fact that there is no common way of even describing curricula. The solution to this problem was a procedure called topic trace mapping, by which researchers in each country collected information about topic coverage in various documents and translated it into a common format. Using formally defined "document analysis procedures" as guides, the national researchers took the most widely used textbooks in their respective countries, as well as national and regional curriculum guides, and analyzed the documents section by section to determine the extent to which material included in the TIMSS frameworks was covered. A total of 491 curriculum guides and 638 textbooks were analyzed. The researchers also asked education experts within each country to respond to questionnaires designed to support the document analyses (Schmidt et al., 1997; TIMSS U.S. National Research Center, 1996).

William Schmidt, who directed the curriculum study, described some of the team's findings and conclusions, focusing primarily on the issues addressed in *A Splintered Vision*, the curriculum analysis results for the United States (Schmidt, McKnight, and Raizen, 1997).[4] For him, the study's most valuable product is what he sees as resolution of the debate over whether school curricula truly make a difference in student learning. For him it is clear that teaching matters, and he argues that the "somewhat disappointing" achievement results for the United States reflect the weaknesses in the U.S. curricula. His conclusion is that many other factors—such as length of time spent in school and assignment of homework—that have been blamed for poor student performance in the United States are side issues. He explained that his research has shown that "there is a tremendous amount of variability across these countries in terms of the way in which mathematics or science is taught." He suggested that further exploration of the relationship between achievement and topic coverage in the curriculum will clarify the picture of student learning considerably.

Specifically, Schmidt argued that no intellectually coherent vision guides mathematics and science curriculum development in the United States. Because responsibility for curriculum decisions rests with states and localities, there is variation among the curricula used within U.S. borders, just as there is among those of different nations. Some of this variation reflects differing educational goals and philosophies, while some of it is, in effect, coincidental. Schmidt presented a few specific findings to illustrate his points:

- Both science and mathematics textbooks in the United States include far more topics than was typical for other countries at all three grade levels. This is true even for science texts devoted to particular topics, such as earth science or physical science.
- Mathematics curricula in the United States consistently cover far more topics than is typical in other countries. In science, the tendency toward breadth is similar, though less pronounced.
- Topics remain in both the mathematics and science curricula for more years in the United States than in all but a few other TIMSS countries. The U.S. practice is to introduce many more topics than do other countries in grades one and two and then to repeat these

---

[4]William Schmidt served as both the principal investigator for the curriculum study and the national coordinator for the U.S. portion of the achievement study. He also served as the project director for the Survey of Mathematics and Science Opportunities (SMSO). This study, conducted in advance of TIMSS, produced a set of classroom observations in six countries that were designed primarily to identify important themes and issues to be explored in the TIMSS background questionnaire. His presentation drew on all of these sources.

topics through grade seven. Schmidt emphasized this point by noting that although new elements called for by science standards have generally been added to the curriculum, little has been removed to make room for the new.

American teachers, Schmidt argued, are sent into their classrooms with a mandate to teach using curricula that reflect few decisions about priorities, are fragmented, and are poorly integrated with one another. Teachers, he said, are armed with textbooks that are similarly laden with a jumble of topics. The curricula are, in his words, "a mile wide and an inch deep." How do teachers handle this situation? Schmidt argued that the instructional decisions made by U.S. teachers mirror the inclusive approach of the tools they are given. Teachers cover more topics, he suggested, but spend less time and emphasis on each than do many of their international counterparts. Instead of "telling a story" about a particular topic, allowing enough time for students to learn it and move on, he argued, U.S. teachers tend to keep reintroducing topics that have not yet been mastered.

Schmidt concluded that the U.S. educational vision is splintered because the U.S. system has many actors and is characterized by "dispersed control," as Richard Elmore later put it. For Schmidt, this system is responsible for the seriously inadequate sets of curricula currently in use. The incoherence of the curricula, he argued, has impeded student learning.

The Three-Country Qualitative Studies

Germany, Japan, and the United States participated in additional studies, sponsored by the United States, in order to augment their understanding of the achievement results. These studies, a videotape analysis and a set of case studies, were devised to explore both instruction and the cultural contexts within which the learning and teaching of mathematics take place. They involved methodologies rarely used in conjunction with large-scale assessments of achievement, and, in the case of the videotape study, of technology developed specifically for TIMSS. James Stigler and Harold Stevenson, the principal researchers for the videotape study and the case studies, respectively, each described their methods and some key findings.

Videotape Study

The primary goal of the videotape study was to capture and then analyze entire mathematics lessons taught to a subsample of the Population 2 (grades seven and eight) students. Lessons were taped in a total of 231 classrooms across the three participating countries. Teachers were asked to make no changes in their normal classroom routines for the videotaping sessions. Standardized camera procedures and other protocols were developed for the data collection. The thousands of hours of tape were digitized, and computer software was developed for analyzing them. Thus it has been possible for researchers to scan quickly through the material on a computer and to search it in various ways.

In addition, the tapes were transcribed and translated and then coded for the occurrence of various events, teaching strategies, and content elements. The coding made it possible for researchers to analyze the lessons quantitatively and to explore such issues as amounts of time spent on seatwork and classwork, discussing and doing homework, and non-lesson activity. The tapes were also analyzed by mathematicians for mathematics content.

In addition to making possible the exploration of questions about teaching practice, as well as specific questions raised by data from the achievement tests and the background questionnaires, the videotapes have two other important uses. First, as symposium participants who watched just a few short segments emphasized, the opportunity to observe a lesson on tape is far more powerful than any verbal description can be. It is clear that the tapes themselves, as well as the experience gained in collecting them, will be an extremely valuable resource for teacher training, as well as for research. Second, the digitized tapes are a permanent, unchanging resource. Future research can be conducted using these tapes as a record of teacher practice at a particular time, as research questions change.

Apart from the interesting technical issues Stigler and his team faced, the videotape study produced some interesting conclusions about variations in teacher practice among the three participating nations. The report on the study had not been released at the time of the symposium, but Stigler discussed several of its key findings. Perhaps most important was Stigler's conclusion that the majority of prescriptions about teacher practice that have been generated by the research community in recent years have not been implemented in U.S. classrooms. Stigler argued that the relatively large-scale videotape study has made it possible for the first time to look at what teachers are actually doing in the classroom and to compare that with their verbal descriptions of what they believe they are doing.

Citing the notion of problem solving, for example, a traditional mathematics skill that is carefully redefined in the National Council of Teachers of Mathematics (NCTM) standards, Stigler pointed out that the understandings teachers and others have of what it means in practice vary to an alarming degree. He described a lesson he had observed, in which students solved a series of traditional word problems as a group. Their teacher had spoken enthusiastically about the "amazing problem solving" the students were doing, believing that she had fully responded to this aspect of the revised standards.

Stigler made the further point that major shifts in education policy often occur without the benefit of any, or sufficient, data about the extent to which the current policy has actually been implemented in the classroom. This point is relevant to a question that many have asked about TIMSS—whether TIMSS achievement results could be seen as a measure of the impact of the NCTM standards, which were published in 1989, on student learning. Stigler's conclusion from the videotapes is that the question is moot since the NCTM reforms have, by and large, not been implemented in U.S. classrooms.

Stigler also reviewed some of what the videotapes revealed about differences among the three nations. He noted that the sample sizes chosen, 100 teachers for Germany, 81 for the United States, and 50 for Japan, partly reflected expectations about how much teaching styles were likely to vary within each country. Surprisingly, Stigler found that teachers in both Japan and the U.S. were remarkably consistent. In general, Stigler's portrait of typical approaches to lessons in Japan and the United States (his presentation focused on these two nations) is likely to cause concern in the U.S. education community, and that impression was strongly reinforced by the videotapes he showed.

The Japanese lesson showed a teacher who pushed his students to grapple with a series of problems and to come up with alternative solutions. The teacher communicated respect for his students' abilities to cope with challenging material, and he guided the students skillfully from the alternative solutions to a more general understanding of the concept the lesson covered. In contrast, the U.S. teacher seemed to lead his students by the hand through an explanation of a concept, and he telegraphed his expectation that the students would have trouble applying the concept in a challenging problem by warning them repeatedly about a particular problem as they began their seatwork. Then, before they had had time to attempt that problem, he stopped them and led them through it step by step. The U.S. lesson was also interrupted more than once, both by conversation about school schedules and other issues unrelated to the lesson and by an announcement over the public address system.

These two excerpts were chosen by Stigler to represent what he and his team had judged to be typical of the lessons he saw in the two countries, and they raise issues that are familiar to many in the policy and research communities. For Stigler, the videotapes from Japan and the United States painted a consistent picture of two different approaches to teaching. He noted that the questionnaires administered to the teachers who participated in the videotape study (these were different from the questionnaires administered with the achievement tests) revealed very different expectations for the outcome of a lesson: 70 percent of Japanese teachers reported that their goal was to get the students to understand a concept; similar percentages of U.S. (and German) teachers reported a goal of getting students to be able to do a certain kind of problem. In Stigler's view, the Japanese lessons generally "tell a story" and provide students with the opportunity to struggle with and explore the concept the teacher is presenting. In contrast, the videotapes show relatively less development of concepts in the U.S. lessons, which Stigler characterized as focusing on short-term goals.

To support his conclusions, Stigler explained a few of his specific findings:

- While the proportions of time spent on individual work and work as a class are roughly the same in the two countries, Japanese teachers tend to switch between the two much more frequently than do U.S. teachers.
- The U.S. teachers pay far more attention to homework than do their Japanese counterparts, allotting significant chunks of class time for going over previous homework or allowing students to begin new assignments, leaving relatively less time for instruction.
- The U.S. lessons were interrupted by non-mathematics-related activities significantly more frequently than were the Japanese lessons. This finding reinforced for Stigler the sense that in Japanese society the lesson is regarded as a coherent, sustained inquiry into a topic while in the U.S. it is regarded more as an episode or a practice session.
- Japanese teachers generally focus on just one topic during each lesson; U.S. teachers average close to two topics.

The participants' responses to the brief videotape excerpts were extremely lively, and many remarked on how convincingly the excerpts seemed to illustrate particular arguments about teaching practice. Some of the issues raised both in the papers prepared for the symposium and by participants about ways of using and understanding this kind of data are explored below ("Critiques and Methodological Issues").

Case Studies

While the primary focus of the videotape study was on teacher practice, the case studies conducted by Harold Stevenson explored in detail the contexts that shape the experiences of students and teachers. Like other parts of TIMSS, this study was designed to provide data to help account for some of the variations in student performance, in this case by examining contextual influences. Previous studies have shown that differences in curriculum and education structure can provide insights about performance, but other kinds of information are also needed. How do teachers in different places think about teaching, learning, and curriculum? How have they been prepared and what kinds of support do they receive? What factors in and out of school affect students' motivation to learn? What are students' attitudes about mathematics and its value?

While the contexts that shape learning can be explored through written questionnaires, the case studies were an opportunity to make cross-cultural comparisons in far greater detail and to investigate subtler issues than a coded questionnaire could permit. Through this project researchers intended to produce thorough analyses—case studies—of education-related factors in three distinct cultures, the United States,

Germany, and Japan. The studies were structured around three basic issues. The first was how content and performance standards in the three countries compare. Through this comparison, researchers hoped to explore the ways in which each country deals with individual differences among students. The second was the role of school in adolescents' lives. The last was the ways that training, certification, and support for teachers' continuing professional development affect their working lives.

Stevenson began by explaining that the study was a sort of hybrid, devised for TIMSS, between the methods of anthropological ethnography and the interview approach characteristic of psychology. The result was what he termed "a descriptive study"—"a description of what you would find if you were in these particular cultures." The basic plan was to identify and train individuals who were familiar with each of the three societies, fluent in the requisite language, and skilled in observation and interview techniques, and to send them into the field to collect information. With the help of country experts, sites were chosen that would broadly reflect national characteristics, and researchers assigned to each country spent 2-3 months collecting data.

The researchers spent the bulk of their time interviewing parents, students, and teachers and observing classroom lessons. They visited homes, schools, and education ministries. The result was hundreds of hours of audiotape, which was transcribed and translated. As in the videotape study, the material was entered into a computer and coded so that researchers could search it efficiently, but the data were not analyzed statistically; rather, they were synthesized into detailed descriptions, organized around the explicit questions that guided the study.

Like the report on the videotape study, the case study reports had not been released at the time of the symposium, but Stevenson highlighted some of the insights that have emerged. One important focus of his presentation was on ways in which detailed knowledge of cultural contexts can significantly alter discussions about a particular issue. His choice of an example—homework—was inspired by his concerns about the ways in which symposium participants had discussed the relationship between homework and achievement results. He noted that in Japan there are four possible translations for the term, none of which corresponds to our notion of the word. The Japanese terms describe a variety of activities one might do outside of class—study, work on practice questions, or do an assignment, for example. They reveal that ways of categorizing such activity differ in the two cultures. To further illustrate the point, Stevenson noted that the amount of homework done by German students varies significantly, depending on the type of school they attend. Consequently, a mean for homework done in Germany would have very little value. It is only through interviews, Stevenson maintained, that researchers were able to discover what kinds of out-of-school studying students in each of the three cultures did. Lois Peak of NCES later made the point that the written questionnaires used carefully chosen language to ask about homework because staff were aware of the issue. Nevertheless, Stevenson maintained that far more sense could be made of such an issue through observation and interview than through a questionnaire.

Another example that Stevenson addressed—which had already been raised several times during the day—was that of "juku," the after-school classes attended by many Japanese students. Stevenson's point was that "juku" is a very vague term that refers not only to intense academic classes, but also to craft classes, sports, and other structured social activities. Many U.S. observers have made the claim that the Japanese students' superior performance can be explained by their attendance at juku because they have assumed that it provided students with rigorous training for college entrance exams and would compensate for any weaknesses in the schools' academic programs. Stevenson claimed that a deeper understanding of the cultural context reveals that this is not true, or at least that it is a seriously oversimplified portrayal.

Stevenson described a few other findings from the study:

- The role of the school principal in Japan is very different from that of one in the United States. In Japan, committees of teachers have primary responsibility for running the school; the principal serves primarily to "execute" the committee's decisions.
- Classifications of student ability come at different times in the three countries. In the United States, the urge to assist children who need it often leads to tracking decisions as early as kindergarten. In Germany, a formal decision is made at the end of fourth grade. In Japan such evaluations are made much later.
- The Japanese curriculum is "a set of broad guidelines of the kinds of things that should be accomplished at each grade level." Teachers are then given considerable latitude to develop specific expectations for different children. In Germany, Stevenson found, the situation is more similar to that of the United States in that each state is empowered to adopt its own guidelines. The German states are, however, required to meet broad national guidelines.

To provide a sense of the flavor of some of the material the study produced, Stevenson read extended quotations from several teachers. He closed by remarking that "it is these kinds of . . . vivid, vital responses that we think give a meaning to a case study . . . that is very difficult to come up with in any other way."

CRITIQUES AND METHODOLOGICAL ISSUES

Lynn Paine, one of the session moderators, expressed a key issue facing the participants when she pointed out that they had been shown graphs of international achievement scores for thousands of students in the morning and a videotape of "one classroom, one teacher, a small number of students" in the afternoon. "How," she asked, "do we somehow bring those together?" Her question reflected not despair but a sense that the challenge presented by TIMSS is a new one. As was repeatedly pointed out, TIMSS includes data drawn from different samples and by means of different methods; moreover, the two three-country studies were added to the original TIMSS design (at the urging of the United States), and there is no detailed blueprint for fitting these elements together.

Clearly TIMSS offers risks as well as possibilities. As one of the symposium paper authors, Michael Huberman (1977:1), wrote: "Such a study could run the risk of the centipede, marching off in several directions at once." The results available so far suggest that different, and possibly conflicting, conclusions might be supported by different parts of the study. Moreover, because the qualitative studies are innovations, neither means of verifying their results nor standards for evaluating their methods are readily available. This section explores questions raised about aspects of the study and the larger issue of linking its components.

Linking the Components of TIMSS

A certain amount of ambiguity may be an inevitable outcome of a study so large and complex. Theoretical or political concerns may drive observers to focus more closely on either the implications for curriculum raised by Schmidt's work or the concerns about teacher preparation raised by Stevenson, for example, given that the study itself was not designed to indicate which finding deserves more weight. For purely practical reasons, few observers may have both the time and skill to truly digest all that TIMSS has to offer. This point need not diminish the usefulness of the study's component parts, but it will surely affect attempts to integrate them.

Nevertheless, the study components each make a contribution to answering core questions about teaching and learning in mathematics or science, and they should be considered as a package. At the time of the symposium, the first TIMSS reports had just been released, and it will not be until some time in 1998 that the last of the reports documenting the primary analysis for each of the study components will be released. Links among the components of the study were not really forged during this first stage. However, the ways in which these links are forged once the primary analyses are completed will be crucial, and symposium participants stressed the importance of establishing a clear linking framework. Several key points about the links were made at the symposium:

- *For the components of this study to be effectively linked, relationships among different research disciplines will need to be estab-*

lished. Scholarly communities that are not accustomed to working with one another's data will need to collaborate in innovative ways to make the best use of the findings from TIMSS.

- *What happens with TIMSS will be a model for the future.* Lois Peak reported that NCES is considering using videotapes in future studies, but she noted that using this powerful tool in valid ways is not a straightforward task. Given the initial reaction to what is known about the qualitative studies and the publicity they have received, it is likely that other researchers are already considering applying these methods in other contexts. The education community has a considerable appetite for rich data about teaching and learning, but, as many at the symposium pointed out, these new kinds of data can easily be misused.
- *Simplistic understandings of TIMSS may be misleading.* Until the links are forged and subjected to rigorous scholarly scrutiny, there is a danger that observers will use "common sense" to link the data from the various components of TIMSS, perhaps yielding misleading results. Observers who do not pay close attention might easily miss the fine points in this complex study—the fact that some data comes from only 3 nations and some comes from 41or 26, for example—and make erroneous conclusions about explanations for achievement results.[5] There are obviously many other differences among the study's components that are salient to any analysis that draws on more than one.

The Achievement Study

As has been noted, many presenters marveled at the magnitude of what TIMSS accomplished. One described it as "a researcher's treasure trove," and many noted that analyses using the data could easily occupy the research community for many years. However, since the achievement component of TIMSS is the base on which the study rests, it is worth noting that several presenters expressed caveats about it. Jan de Lange, noting that multiple-choice items have been outlawed in his country, The Netherlands, argued that the TIMSS items are primarily useful for testing low-level knowledge and do not necessarily represent anyone's idea of a desirable curriculum. In their paper, Atkin and Black (1997) expressed a similar concern, noting, for example, that a total of 11 multiple-choice and 3 free-response items were used to test the middle school population's knowledge of

[5]Population 2 students in six nations were surveyed in the Survey of Mathematics and Science Opportunities. The topic trace mapping components of the curriculum study covered 46 nations, and that study's survey of teachers covered Population 2 students in three nations. The videotape study and case studies each involved only Population 2 students in Germany, Japan, and the United States. Finally, as noted, the achievement results were reported for Population 1 students in 26 countries, Population 2 students in 41 countries, and Population 3 students in 21 countries.

the portion of the test domain identified as "Environmental Issues and the Nature of Science." First, they argued, from this "small number of questions the results can hardly be a substantial basis for firm conclusions." They also noted that these 11 questions cover two distinct content areas, whose relationship to one another is not explained in the framework (Atkin and Black, 1997:12-13).

Others made similar comments, but most, including de Lange as well as Atkin and Black, acknowledged that it would likely not have been possible to conduct the assessment at all without using methods that are both efficient and well established. Nevertheless, participants noted how easy it is for observers to lose sight of exactly what was assessed as the results are disseminated and applied in various contexts.

The Curriculum Study

A number of participants raised questions about the curriculum study, primarily focusing on the conclusions Schmidt drew from his findings. For example, several questions focused on what TIMSS suggests about the ways that control over education systems might interact with achievement. In response to Schmidt's argument that U.S. students' relatively low performance is the result of an incoherent curriculum, Atkin and Black made reference to results indicating that TIMSS does not reveal a clear correspondence between centrally controlled, and, by implication, coherent, education systems and achievement. Schmidt responded by noting that even a very focused curriculum may not be implemented in the classroom in a coherent manner.

Others raised questions about whether the available means of measuring and comparing curricula were truly sophisticated enough to support the detailed comparisons that have been made. Still others pursued this point from a different angle, questioning whether the impact of the structure of curricula and textbooks can be isolated as a factor, separate from the ways they are translated into classroom instruction. Schmidt argued that it can, though he noted that U.S. curricula and textbooks may not be functioning as they are intended to. For example, he explained, textbook publishers have made rational marketing decisions in choosing to reflect a variety of curricula in their books. Their intention has been that teachers will use only the material that is relevant to the curricula they are following. Schmidt's point was that if the system is not working, only systemic changes can effectively improve student performance. "The problem," he maintained, "is in the curriculum policy area, and the only way it can be addressed is . . . as a nation."

The Qualitative Studies

Another issue presented by TIMSS is that both of the qualitative studies took existing methods and "ratcheted them up," in the words of one participant, to new levels of both scale and sophistication. Before even addressing the links among them and the achievement and curriculum data, observers have begun to assess these studies themselves. Not surprisingly, because of its novelty, the videotape study dominated the discussion.

Michael Huberman raised several important issues. He offered a general critique of the study's theoretical underpinning (see below, "Policy Issues"), but he also raised some specific questions about the methods of the videotape study. First, he pointed out, although the videotape certainly provides a far more detailed picture of the classroom than questionnaire data could possibly have done, the picture is still far from complete. Students and school culture, for example, contribute a great deal to the nature of a classroom lesson and have considerable influence on teachers' decisions, both large and small. A videotaped lesson, Huberman argued, is not easy to interpret in the absence of knowledge of its context. An understanding of what occurred during the days preceding and following the lesson that was videotaped might significantly alter an observer's interpretation of the lesson.

A related issue for Huberman was that the videotapes provide a very "teacher-centered" vision of the lesson. They cannot reveal how students have perceived the lesson. Researchers coded teacher responses for "helpfulness" as part of their analysis, for example, although they had no means of knowing whether students had perceived that they had been helped by the interaction in question.

The coding was also an issue for Huberman for another reason. What, he wondered, is the value of collecting data as rich as these videotapes, and then immediately coding it and reducing it to statistics that can be put into tables? Moreover, he asks, is there not a danger in the "irresistible analytic convenience" of the software? Might not the software's power in counting the frequency with which certain behaviors occurred have "tricked" researchers into "unearthing 'themes' or 'patterns'" that were not actually there (Huberman, 1997:14)?

Huberman also raised questions about the sampling for the study. Pointing out that the sampling was not random, Huberman noted in particular that the three types of schools in Germany, the hauptschule, the realschule, and the gymnasium, which differ in significant ways, were not represented proportionally. He also raised a question about how the high refusal rate (almost 50 percent) among schools that were asked to participate might have affected the outcome. Although the study included a record number of classrooms, it nevertheless runs the risk of seeming to be no more than an unusually rich collection of persuasive anecdotes.

Huberman also noted that the effect of the cameras on the teachers and students who were filmed could not be known. Stigler had addressed that issue in his presentation because it had been an important concern for his team. Their conclusion was that while teachers' and students' awareness of the camera may have affected their behavior in a variety of ways, it is not likely that teachers could actually change their teaching in fundamental ways likely to alter the study's results. If they could, Stigler joked, the installation of cameras in classrooms would be a simple means of improving teaching.

A final set of questions Huberman raised concerned the fact that the study filmed a single 50-minute lesson in each of 231 classrooms. Huberman wondered whether filming a series of lessons in a smaller number of classrooms might have yielded more useful results. Many of the coding categories, he noted, were efforts to capture "activities or processes that play out over time," such as building on complex concepts or establishing links with content covered previously, that can not easily be evaluated in the context of a single lesson (Huberman, 1997:12).

Although Huberman's primary contribution was to raise questions about the study, he nevertheless described it as an extremely impressive effort. Symposium participants did not have sufficient time to wrestle with all of the questions, or to resolve any of them, but they did refer to many of them in various contexts. After watching two excerpts from the videotapes, participants also raised another concern. As was discussed above, many at the symposium had enthusiastic reactions to the videotapes and launched eagerly into discussions of what the lessons shown demonstrated. But as Lois Peak pointed out, the powerful reactions people had illustrate the risk that the videotapes could be misused: because they are so much richer and more compelling than written descriptions, viewers may feel a sense of certainty about impressions based on them that is unwarranted.

This richness is, of course, their virtue as well. Lynn Paine cited as an example of this something she observed in the two lessons that were shown. Both could be described as decidedly teacher directed, but their ways of being so were dramatically different. In the U.S. lesson, she pointed out, the teacher was evidently perceived as the sole source of both information and ideas; students in the class did not look at others who were speaking, or seem to engage as a team. In contrast, the Japanese teacher had clearly planned the lesson around the idea that different students would come up with different valid means of solving problems. He showed that he intended the students to learn from one another as well as from him, even though he retained control of the discussion.

Part of Paine's point was that this sort of insight is valuable regardless of how representative a particular lesson or behavior might be. In a larger sense, this point applies to many aspects of TIMSS. While forging links among the components will be extremely important, the separate sets of data can be of significant value on their own to both policy makers and others who are seeking to evaluate policies and strategies, and to practitioners who are seeking insights or inspirations. TIMSS is not a research project designed to test pre-existing hypotheses, as Edward Haertel pointed out; its results cannot be used to conclusively prove or disprove assertions. It provides no control groups because the context in which each participating student has learned science and mathematics is different. However, clear evidence that a particular intervention had a particular result is not necessary to make the data useful. Data from each of the components can be used to enrich understanding of education and, more significantly, to identify promising connections that can then be further explored.

POLICY ISSUES

The many policy issues raised by the initial findings from TIMSS were on the minds of presenters and participants alike throughout the symposium, and many of them were raised more than once in different contexts. The issues raised can perhaps most easily be summarized as four basic messages that were drawn from what was known about TIMSS at the time of the symposium.

Understanding Differences Among Countries

The TIMSS results clearly highlight the importance of understanding differences among countries. This issue, while seemingly obvious in that the purpose of TIMSS is to compare the educational structure and performance of participating nations, was manifested in two particular ways at the workshop that will be of interest to policy makers. The first of these was primarily addressed by Mike Atkin and Paul Black, whose paper summarized some of the results of a 13-country study, the Innovations in Science, Mathematics, and Technology Education Project, sponsored by the Organization for Economic Co-operation and Development (OECD), for which they collected case studies of innovative approaches to mathematics and science education (Atkin and Black, 1997).[6] From this work they concluded that while every single participating nation (including those that performed well on TIMSS) is decidedly dissatisfied with the status of its own approach to mathematics and science education, not all nations share the same motivations for seeking improvement. Many, particularly those facing high unemployment, share with the United States an overriding concern with preparing young people for the labor market and using a focus on excellence in mathematics and science as a means of improving productivity and fostering economic growth. Others were motivated by quite different concerns, such as the state of adolescent health, or the need to address environmental deterioration (Atkin and Black, 1997:5). According to Atkin and Black, Japan is primarily motivated by the concern that its students are not suffi-

[6]The 13 countries involved in the study were Australia (Tasmania), Austria, Canada (British Columbia and Ontario), France, Germany, Ireland, Japan, Netherlands, Norway, the United Kingdom (Scotland), Spain, Switzerland, and the United States.

ciently creative, despite measurably high achievement, and their reforms have been generally targeted toward fostering innovative problem-solving skills and encouraging real-world applications for mathematics and science education.

This point was echoed by Jan de Lange, who cautioned that to an observer from abroad, the United States' virtual obsession with economic competition, particularly with Japan, is, to say the least, puzzling. He reminded participants that somewhat loftier goals for education—the proposition that "it makes people richer intellectually and culturally and prepares them for an increasingly complex society," for example—have a practical application (de Lange, 1997:7). Such goals, he argued, can enhance the development of intellectually rich academic standards that are appropriate to their context. He suggested that the heavy emphasis on standardized test scores in the United States has distorted both curricula and expectations for student learning. Atkin and Black made a similar comment, noting that "there is no substitute for hard argument within each country, to formulate the standards of high quality that it values and to work out the policies that can help achieve those standards" (Atkin and Black, 1997:16).

Atkin and Black stressed that their experience with the OECD study makes clear that the TIMSS results are a snapshot taken at a fixed point in time—a snapshot of student performance and of educational systems that are in near-constant flux. Their point, that the TIMSS results must be seen as a baseline against which changes in education can be marked, was shared by symposium presenter Richard Elmore, who demonstrated a second reason that the context for each country's performance is so crucial. Elmore's focus was on the role TIMSS plays in the education policy environment in the United States, and his argument was that the study provides a unique opportunity in this country because of the time at which it was done (Elmore, 1997). This, he argued, is a time when the proposition that imposing formal standards for students, teachers, and schools has real potential for improving U.S. schools has achieved an almost unprecedented level of agreement among concerned groups. Consequently, he maintained, the data produced by TIMSS, which includes detailed information about classroom practice, curriculum, teacher preparation, and many other contextual factors, should provide support for education leaders who want to take standards the crucial step forward, into classroom practice.

Elmore structured his argument around a description of the U.S. political system as being characterized by both pluralism and dispersed control. Tying these characteristics to our education system, Elmore pointed out that the system is pluralistic in the sense that any constituency that is able to muster a critical mass of support can have an impact on education policy. He argued that the structure of education governance in the United States is neither centralized nor, though it is often so described, localized. Elmore prefers to describe control over education governance as "dispersed": depending on the power of interested constituencies, influence can be wielded at any level. Though central controls are not prevalent, he noted, the federal government intervened with force in support of school desegregation during the 1960s. More typical are situations in which constituencies with differing views seek in their own ways to influence policy decisions made at various levels, and the outcome is determined largely by political clout. It is because of this possibly unique system that the current apparent consensus over the value of education standards is so remarkable, said Elmore.

Typically, Elmore argued, the dual effects of pluralism and dispersed control have helped to ensure that most "policy talk" is carried on at an abstract level and has little impact on the day-to-day negotiations about specific decisions. (Elmore credited Tyack and Cuban, 1995, for this point.) TIMSS presents the novel possibility that policy prescription could move into the "instructional core," as he put it, by influencing decisions about "what gets taught to whom." TIMSS was designed to investigate the links between achievement and contextual factors and was based on the conviction that classroom decisions and other contextual variables have significant effects on student learning. For this reason, Elmore argued, it should provide real support for policy decisions that truly confront what are for him the two key issues for the success of the standards movement, capacity and incentives.

Elmore formulated what he described as a new principle, "reciprocity of capacity and accountability," to explain his conception of how standards-based reform ought to proceed. His concern is that holding schools accountable for student performance is tremendously risky (Elmore, 1997:15):

> Race, social class, and home environment are the strongest predictors of education performance for students. Rewarding and punishing schools based on their performance under these circumstances means rewarding and punishing them, in effect, for the students they serve. Worse yet, adjusting rewards and punishments for student background means that certain schools will be allowed to continue to have lower expectations for their students than other schools, thus defeating the main purpose of standards-based reform.

Acknowledging that this is, as he put it, "a horrendously difficult problem," Elmore maintained that TIMSS can play a valuable role in focusing discussion on the issue. The study strongly emphasizes the connection between student learning and the many influences on teachers and schools that affect it. Consequently, it supports his argument that identifying and providing the supports necessary to enable students, teachers, and schools to meet established standards will be crucial to the success of standards-based reform.

Jan de Lange had a somewhat different perspective on the same issue. He noted that "there is no mechanism that steers innovation in the United States." He added that although the United States spends more money than any other country in the world on research about mathematics education, to an outsider it does not seem that this research has provided as much benefit to students and teachers as it should have. Because so many decisions about school governance are made at the local level and because, he said, "the school board people are not always, let me put it gently, experts in education" he believes they are not particularly likely to be aware of, or persuaded by, education research. De Lange and Elmore shared a conviction that for improvement to occur, the gap between research and theory, on one hand, and practice, on the other, must be bridged.

Finally, de Lange reiterated the point that understanding of the contexts that influence education within each country is indispensable. He called for a focus on variations of performance within nations as well as those between them. Citing the vast differences that have been revealed (through the International Assessment of Educational Progress) between the performance of students in Iowa, North Dakota, and Minnesota and that of students in Alabama and Louisiana, for example, he remarked that "this gives at least a suspicion that we cannot blame textbooks or curriculum alone. He maintained that this variation in performance ought to be "unacceptable" (de Lange, 1997:10).

Support for Teachers

Although discussions throughout the symposium touched on issues that revealed potential conflicts of various sorts, two basic points of agreement emerged clearly. Perhaps clearest was a ringing endorsement for the idea that teachers in the United States require far more support than they are currently getting if they are to effect the desired improvements.

Jan de Lange remarked that he had "never seen teachers working under [such] bad conditions . . . as American teachers" and deemed it "remarkable that we still end up in the middle" under these circumstances. He cited their few opportunities for professional development, their low status, and the incoherence of the system in which they function as just a few among the many problems they face. Mary Lindquist followed up by noting that in her experience working with teachers, what they want most is "the time to do the things that they think they should be doing."

Atkin and Black addressed the role of teachers from a different angle. One of the conclusions they drew from the OECD project was that the absolute dominance that university-based scientists and mathematicians have had over the content of K-12 instruction is declining. Teachers in particular, they noted, are gaining new influence in determining what should be taught, at least in some areas. However, as they put it, "change creates turbulence" (Atkin and Black, 1997:11). For teachers to exercise this influence comfortably, Atkin and Black explained, they need opportunities for collaborating with their peers, and for upgrading and maintaining their own subject knowledge. They called attention to some revealing data from TIMSS showing that U.S. science teachers average significantly fewer hours per week devoted to both professional reading and development and to lesson planning than did such high-scoring countries as Japan, Hungary, and Singapore (Atkin and Black, 1997:13).

Elmore also addressed the urgency of attending to what teachers need in order to do their jobs well. He noted that "the work day of most teachers is organized in a way that allows them virtually no time to engage in any sustained learning about how to do their work differently," and that "most professionals learn new practices by working with other professionals, in close proximity to the details of practice, and by making their clients pay for the surplus time required to retool and renew themselves" (Elmore, 1997:13). He views it as critical that teachers be given similar opportunities at the same time they are required to meet new standards.

He also noted how ill-suited most existing standards documents are for helping teachers make immediate decisions about what and how to teach. To be useful to teachers, he argued, these documents need to take account of the lesson time teachers actually have and to be "drastically pared, simplified, and operationalized in the form of lesson plans, materials, and practical ideas about teaching practice" (Elmore, 1997:12). In general, participants and presenters clearly seemed to agree that while teachers need to be held to high standards themselves and to significantly raise their expectations for U.S. students, they need to be supported in doing so with concrete and well-planned allocations of time and training.

Secondary Analyses of TIMSS Data

The other basic point of agreement at the symposium was that, despite numerous cautions and criticisms, the TIMSS data are extremely valuable and can serve as the platform from which a wide variety of secondary analyses can take off. The bulk of the specific suggestions for valuable secondary analyses based on TIMSS data came from Edward Haertel, who had been asked to discuss the issue at the symposium. He began with the premise that linking single variables to achievement would likely be unprofitable. "The answers to all such questions," he wrote, "are likely to be equivocal, with many factors each being found to matter a little" (Haertel, 1997:5). For example, he explained, "more than two hours per day of television viewing may be associated with lower achievement, but it does not follow that [students'] watching less television will cause achievement to rise." He also noted that it may be far easier to use TIMSS data to identify factors that have no apparent effect than to calibrate the relative effects of those that are influential.

Haertel's suggestion for approaches to more fine-grained analyses of the data is to break them down in various ways. By exploring subsets of test questions, or items, he explained, it should be possible to begin addressing in detail some questions with significant policy implications. Clusters of items could be defined in a variety of ways—for example, by mathematics or science topics or by the type of task the item calls for. Alternatively, clusters of students could be defined by demographic factors, by exposure to particular material, or by school characteristics. Another approach would be to select subsets of items by statistical characteristics and then try to determine whether they share any features. Generally, looking at targeted portions of the data could provide answers to specific questions about the relative effects of various factors on achievement.

Haertel pointed out that scores varied far more within individual nations than they did between nations, and he said that gaining understanding of reasons for this would be extremely useful. The United States, he noted, has the third greatest variation in scores of the 41 nations that participated in the middle-school (Population 2) portion of TIMSS.[7] One constructive response to that fact, he argued, would be to try to learn from the exceptions, to ask: "Where do the poor learn as much as the wealthy? . . . Where are classes large and resources meager but achievement still high?"

Haertel encouraged observers who are not psychometricians to participate in the formulation of questions to be addressed using the TIMSS data. He suggested four examples of areas of policy interest that could be explored, while acknowledging that there are many others:

- What are the patterns of gender differences in mathematics and science achievement in different nations?
- How does the variability in educational opportunity and outcomes within the United States compare with that within other nations?
- How widely are new ideas about mathematics curriculum and instruction being implemented?
- Do new approaches in instruction, school governance, or other areas, seem to lead to distinctive patterns of student achievement?

In general, Haertel suggested, the cross-national comparisons made possible by TIMSS are "sources of hypotheses of what to look for within the United States." Specific hypotheses cannot be tested using TIMSS data alone, he noted; the national populations are not comparable, so evidence of success with a particular approach in one place cannot be transferred to another. Haertel offered a reminder that TIMSS is not an instrument for comparing the results of educational "experiments" conducted in "laboratories" around the world, but a

[7] Among the participating nations, the standard deviation ranges from 72 to 111; the standard deviation for the U.S. science scores is 106. The standard deviation of the national averages is approximately 50.

comparative observational study. "The most powerful uses of TIMSS," he explained, "may be to show us the range of the possible."

Limitations of TIMSS

Symposium presenters were perhaps most outspoken in describing some of the "yellow lights" they wanted to hold up about ways in which the TIMSS results might be used or misused. Foremost among these concerns was that the study and its results are complex and that it is very tempting to oversimplify them in talking about their implications. Participants emphasized their concern, for example, that results from one of the three-country studies of middle-school students might easily be misconstrued as explaining achievement results for the 41 countries that tested that population, or those at the other two age levels.

Another danger of oversimplification was supplied by Atkin and Black, who noted that the practices the education community considers desirable are by no means always characteristic of the countries who performed well. "If . . . the cost of high scores is to incur or exacerbate weaknesses on other important criteria," they explained, "then there [would be] some difficult decisions to be made" (Atkin and Black, 1997:14).[8]

Many presenters and participants also pointed out that the education community actually knows very little about some of the high-performing countries. Since Singapore performed so well, they argued, the next step is to learn more about how that country actually educates its children, rather than to blindly imitate what is already known or, worse, assumed.

A second concern that was expressed by several participants is that TIMSS, although an exemplary assessment by many criteria, is in no way suited for use as a benchmark of world-class performance. As has been noted, the framework on which the achievement results are based covers only the content which the 45 participating countries could agree merited assessment. It does not represent anyone's idea of a valid program of instruction in itself. It is not correct, as Richard Elmore emphasized, that "since the TIMSS study embodies standards that somehow these standards have some sort of authoritative standing as a consequence of having been connected up with very fine state-of-the-art empirical research." Moreover, as Jan de Lange and others made clear, the testing instrument, which had to be both affordable and understandable in countries all over the world, was capable of measuring only a limited universe of material. It was

---

[8]Their point is reinforced by the fourth-grade results, released after the symposium, in which U.S. students at that level ranked considerably higher relative to their international counterparts than did U.S. eighth-graders. Clearly, policy prescriptions designed to make the U.S. system more like those of particular high-performing countries look even less sensible in light of this difference between the grades.

not designed to assess many of the skills identified in current mathematics and science standards, for example, because these cannot realistically be assessed in a large-scale assessment format.

Michael Huberman offered another perspective on the notion of TIMSS as an international benchmark. "There seems to be a *Zeitgeist* permeating the study," he wrote. (Huberman, 1997:7) His suggestion was that the U.S. NCTM standards had a heavy influence on the content framework, and that a policy perspective supportive of national curricula and of a "back to the basics" approach to standards seemed to lie behind some of the decisions about the structure of TIMSS. His concern was that these unexamined assumptions have guided the study itself and will guide interpretation and application of the findings.

Finally, many of the presentations offered reminders that the TIMSS results are not yet fully digested, and are by no means conclusive. Taking as an example the question of a national curriculum, it is clear that many perspectives are coexisting under the tent of TIMSS. The conclusion drawn by Bill Schmidt, based on his study of curricula and texts, is clearly that U.S. students don't perform as well as they could because their instruction is neither coherent nor consistent. While none of the other TIMSS researchers made causal claims as specific, it is clear that other plausible explanations deserve exploration. The preliminary findings from both of the qualitative studies presented at the symposium, for example, highlight compelling observations about classroom practice and contextual factors that might have large effects on student learning.

Atkin and Black clearly took issue with Schmidt's claim that a lack of curricular coherence accounts for the performance of U.S. students, noting that, "there is no strong evidence from the TIMSS data that the existence or absence of a nationally prescribed curriculum leads to improved performance" (Atkin and Black, 1997:15). They noted that "although eight of the top ten countries [in science] all have national curricula, so do eight of the bottom ten," and the results are similar for mathematics (Atkin and Black, 1997:15). Paul Black concluded his remarks with a gloomy scenario for the United States related to this point. "My nightmare," he explained, "is that an incorrect conclusion from the TIMSS data is you need a firm national curriculum [and that] you need regular testing. It has got to be affordable; therefore it will be short; and we have got to do this quickly." Reminding participants that his own country, Great Britain, has recently instituted a national curriculum, Black argued that that experience had yielded little improvement and had damaged teacher morale.

Elmore implicitly addressed Schmidt's call for coherence in America's curricula by arguing, in effect, that it is not politically realistic. Noting that "the temporary bi-partisan consensus on goals and standards that followed from the Charlottesville summit [on education issues] concealed, it turns out, a deep and roiling suspicion of anything 'national' or 'federal' in matters of curriculum and student learning"

(Elmore, 1997:5). Elmore argued that high academic standards can be established, and be effective, without national consensus on precisely what they contain. Since it is states that hold the constitutional responsibility for ensuring that children are educated, it is they who will exert the pressure that will make standards a reality. Elmore's concern is not so much that standards and curricula will not be sufficiently coherent but that if equal attention is not paid to ensuring that supports are in place to assist schools and students that need it, the consequence will be penalties for schools that serve needy populations, and a decrease in the already elusive equality of opportunity that has been a guiding goal for the U.S. educational system.

Jan de Lange perhaps summed up the views of many of the presenters and participants with the following advice, which he addressed to teachers but which could certainly apply more broadly: "Make no changes if not sure of direction."

SUMMARY

The purpose of the symposium was neither to achieve consensus on any of the issues raised by TIMSS nor to formulate specific advice or suggestions for those using the data. Rather, the purpose was to bring together a variety of perspectives in order to stimulate ideas and raise questions. This is precisely what was accomplished, as symposium chair Richard Shavelson noted when he began his summary with the remark that "multiple perspectives prevail." He also noted, however, that "this is a tough message to give policy makers." Despite the fact that discussion and analysis of the TIMSS results are only beginning and that the results so far available have not yielded obvious policy prescriptions, Shavelson continued, several useful themes and questions emerged from the discussion.

Context matters. There was a sense, seemingly shared by virtually all who spoke at the symposium, that student and school performance must be understood in context. As Shavelson put it, "The policy implication is that focusing education reform solely on the schoolhouse and not family, community, and other socioeconomic supports is likely to fall short of the mark." The study was designed to explore both achievement and at least some of the many contextual factors that affect it. The next task, participants seemed to agree, is to ensure that the importance of the relationship between these two is understood as the TIMSS results are disseminated.

Given the importance of understanding each country's results in context, how can the research and policy communities generalize from what TIMSS has shown? Shavelson noted that there is a two-fold issue in this question. It is important, first, to confirm that what appears to be characteristic of a particular country is indeed so—that the data are accurately modeled. Second, context notwithstanding, those interested in TIMSS will want to derive guidance from the study. Acknowledging that specific claims about causation cannot be supported by data from TIMSS, Shavelson went on to argue that responsible uses can be made of the study's results. "How," he asked, "can we account for the particularity of context while reaching generalizations with TIMSS?" While he had no ready answer, he urged the group as a scientific community to continue to address the problem in order to profit fully from the vast investment of money and effort that has been made in collecting the data.

TIMSS provides valuable images of what is possible. Recognizing that the TIMSS results will be generalized by policy makers, educators, and the public, and cognizant of the need for caution in interpreting and learning from TIMSS, Shavelson was enthusiastic about learning from the alternatives TIMSS provides—images of what is possible. These images, he explained, particularly of teaching and of curriculum, can stimulate thinking, provoke public debate, and provide valuable perspectives, long before they have been scientifically scrutinized. While it is true, he added, that questions about generalizing—about whether a strategy will work in another context or can be effectively adapted by another teacher—may remain unsolved, they need not hamper experimentation. Trying out alternatives suggested by TIMSS will be the key to understanding in which contexts, if any, they will succeed.

There is a clear need for ongoing study. The symposium discussion made clear that researchers who have not yet had the opportunity to look at this rich dataset will bring alternative perspectives, and it is important that they gain access to the data. In addition, Shavelson said, the innovative combination of research methods used in TIMSS calls for an innovative combination of researchers to undertake the secondary analysis. A kind of teamwork that has not been tried before may be called for, Shavelson argued, and he urged the community to consider ways of making sure that this happens. He also urged those in a position to do so to feel a responsibility to provide support for TIMSS research beyond what has already been planned and funded. Further research, he argued, ought to represent a diversity of views and to focus on issues that have significant policy implications and, therefore, be useful to policy makers.

TIMSS has some clear implications for education reformers. Shavelson drew from the symposium a clear sense that TIMSS reinforced the notion that no reform ought to be undertaken without a corresponding commitment to do three things: provide adequate resources to support it, sustain it for long enough to be sure it has had a chance to take hold, and evaluate its impact. He expressed a hope that further dialogue and debate based on the TIMSS results would help decision makers focus their reform efforts. Seconding Elmore's view of the political context in which TIMSS was undertaken, Shavelson suggested that the study could help to solidify some of the consensus that seems to be developing around standards-based reform.

REFERENCES

Atkin, J. Myron, and Paul Black

1997 Policy perils of international comparisons: The TIMSS Case. *Phi Delta Kappan* 79(1):22-28.

Bracey, G.W.

1996 International comparisons and the condition of American education. *Educational Researcher* 25(1):5-11.

de Lange, Jan

1997 Looking Through the TIMSS Mirror from a Teaching Angle. Freudenthal Institute, Utrecht University, The Netherlands.

Elmore, R.F.

1997 Education Policy and Practice in the Aftermath of TIMSS. Department of Education, Harvard University.

Haertel, E.H.

1997 Exploring and Explaining U.S. TIMSS Performance. Stanford University.

Huberman, Michael

1997 Qualities of Instruction and Varieties of Qualitative Research. Faculty of Education and Psychology, University of Geneva.

International Association for the Evaluation of Educational Achievement

1996a *Mathematics Achievement in the Middle School Years: IEA's Third International Mathematics and Science Study (TIMSS).* Center for the Study of Testing, Evaluation and Educational Policy. Chestnut Hill, MA: Boston College. Available: http://wwwcsteep.bc.edu/timss [June 19, 1997].

1996b *Science Achievement in the Middle School Years: IEA's Third International Mathematics and Science Study (TIMSS).* Center for the Study of Testing, Evaluation and Educational Policy. Chestnut Hill, MA: Boston College. Available: http://wwwcsteep.bc.edu/timss [June 19, 1997].

McKnight, Curtis C., et al.

1987 *The Underachieving Curriculum: Assessing U.S. School Mathematics from an International Perspective.* University of Illinois at Urbana. Champaign, IL: Stipes Publishing Co.

Medrich, Elliott A., and Jeanne E. Griffith

1992 International Mathematics and Science Assessments: What Have We Learned? Research and development report from the Office of Educational Research and Improvement, NCES 92-011. U.S. Department of Education, Washington, DC.

National Center for Education Statistics

1996 *Pursuing Excellence: A Study of U.S. Eighth-Grade Mathematics and Science Teaching, Learning, Curriculum, and Achievement in International Context.* Washington, DC: U.S. Department of Health and Human Services. Available: http://www.ed.gov/NCES/timss [July 8, 1997].

Robitaille, David F., and Robert A. Garden, eds.

1996 *Research Questions and Study Design.* TIMSS Monograph No. 2. Vancouver, Canada: Pacific Educational Press.

Rotberg, Iris C.

1990 I never promised you first place. *Phi Delta Kappan* 72(4):295-303.

Schmidt, William H., Curtis C. McKnight, and Senta A. Raizen

1997 *A Splintered Vision: An Investigation of U.S. Science and Mathematics Education.* U.S. National Research Center for the Third International Mathematics and Science Study. Dordrecht, The Netherlands; Boston, MA; London, England: Kluwer Academic Publishers.

Schmidt, William H., Curtis C. McKnight, Gilbert A. Valverde, Richard T. Houang, and David E. Wiley

1997 *Many Visions, Many Aims: Volume 1: A Cross-National Investigation of Curricular Intentions in School Mathematics.* Dordrecht, The Netherlands: Kluwer Academic Publishers.

Third International Mathematics and Science Study
1996 *Third International Mathematics and Science Study: Quality Assurance in Data Collection*, M. Martin and I. Mullis, eds. Center for the Study of Testing, Evaluation and Educational Policy. Chestnut Hill, MA: Boston College. Available: http://wwwcsteep.bc.edu/timss [June 30, 1997].

TIMSS U.S. National Research Center
1996 Report No. 7 December 1996. East Lansing, MI: Michigan State University. Available: http://ustimss.msu.edu/nwslet/report7.htm [September 2, 1997].

Tyack, David, and Larry Cuban
1995 *Tinkering Toward Utopia.* Cambridge, MA: Harvard University Press.

Appendix A

Symposium Participants

PRESENTERS

Bruce Alberts, National Academy of Sciences and National Research Council
J. Myron Atkin, School of Education, Stanford University
Deborah Ball, Department of Teacher Education, University of Michigan, Ann Arbor
Albert Beaton, TIMSS Study Center, Boston College
Paul J. Black, School of Education, King's College, London, England
Gail Burrill, National Center for Research in Mathematical Sciences Education, University of Wisconsin, Madison
Daryl Chubin, Education and Human Resources Directorate, Division of Research, Education, and Communication, U.S. National Science Foundation, Arlington, Virginia
Jan de Lange, Freudenthal Institute, Utrecht University, The Netherlands
Richard F. Elmore, Graduate School of Education, Harvard University
Pascal D. Forgione, National Center for Education Statistics, U.S. Department of Education
Edward Haertel, School of Education, Stanford University
Stephen P. Heyneman, Human Resources and Social Policy Division, World Bank, Washington, D.C.
Michael Huberman, Faculty of Education and Psychology, University of Geneva, Switzerland*
Michael Kirst, Advisory Council on Education Statistics, School of Education, Stanford University
Glenda Lappan, Department of Mathematics, Michigan State University, East Lansing

*Unable to attend.

Mary M. Lindquist, School of Education, Columbus State University
Robert Linn, School of Education, University of Colorado, Boulder
Shirley Malcom, Directorate for Education, American Association for the Advancement of Science, Washington, D.C.
Fred Mosteller, Department of Statistics, Harvard University
Lynn Paine, Department of Teacher Education, Michigan State University, East Lansing
Francisco O. Ramirez, School of Education, Stanford University
William Schmidt, College of Education, Michigan State University, East Lansing
Richard Shavelson, School of Education, Stanford University
Harold Stevenson, Department of Psychology, University of Michigan, Ann Arbor
James Stigler, Department of Psychology, University of California, Los Angeles
JoAnne Vasquez, Science/Social Sciences Resource Center, Mesa Public Schools, Arizona
Luther Williams, Office of the Assistant Director, Directorate for Education and Human Resources, U.S. National Science Foundation, Arlington, Virginia

ATTENDEES

Gordon M. Ambach, Council of Chief State School Officers, Washington, D.C.
Karen Anderson, National School Boards Association
Ronald Anson, Office of Education, Research and Information, U.S. Department of Education
Marty Apple, Council of State Scientific Society Presidents, Washington, D.C.
Hyman Bass, Department of Mathematics, Columbia University
Alexandra Beatty, Commission on Behavioral and Social Sciences and Education, National Research Council
Jerry P. Becker, Department of Curriculum and Instruction, Southern Illinois University
Eve Bither, Office of Education, Research and Information, U.S. Department of Education
Erling Boe, Graduate School of Education, University of Pennsylvania
Robert Boruch, Graduate School of Education, University of Pennsylvania
Mary Lyn Bourque, National Assessment Governing Board, Washington, D.C.
Gerald W. Bracey, Consultant, Alexandria, Virginia
John S. Bradley, Division of Elementary, Secondary and Informal Education, Education and Human Resources Directorate, U.S. National Science Foundation, Arlington, Virginia

Sadie Bragg, Borough of Manhattan Community College, The City University of New York

Annora Bryant, Office of Education, Research and Information, U.S. Department of Education

Rodger Bybee, Center for Science, Mathematics, and Engineering Education, National Research Council

Nancy Carson, Nancy Carson Associates, Alexandria, Virginia

Audrey Champagne, School of Education, University at Albany-State University of New York

Richard Clark, Minnesota Department of Education, Minnetonka

Margaret B. Cozzens, OPS Division, Office of the Director, U.S. National Science Foundation, Arlington, Virginia

Mary Crovo, National Assessment Governing Board, Washington, D.C.

Christopher Dede, Division of Elementary, Secondary and Information Education, Education and Human Resources Directorate, U.S. National Science Foundation, Arlington, Virginia

Joan Donahue, National Alliance of State Science and Mathematics Coalitions, Washington, D.C.

Janice M. Earle, Division of Elementary, Secondary and Information Education, Education and Human Resources Directorate, U.S. National Science Foundation, Arlington, Virginia

Emerson Elliott, National Council for the Accreditation of Teacher Education, Washington, D.C.

Joyce B. Evans, Division of Elementary, Secondary and Information Education, Education and Human Resources Directorate, U.S. National Science Foundation, Arlington, Virginia

Lawrence Feinberg, National Assessment Governing Board, Washington, D.C.

Francis Fennell, Department of Education, Western Maryland College

Joan Ferrini-Mundy, Center for Science, Mathematics, and Engineering Education, National Research Council

Michael Feuer, Commission on Behavioral and Social Sciences and Education, National Research Council

Raymond Fields, National Assessment Governing Board, Washington, D.C.

Kevin Finneran, Editor, *Issues in Science and Technology*

Mary Frase, National Center for Education Statistics, U.S. Department of Education

Matthew Gandal, American Federation of Teachers, Washington, D.C.

Alice Gill, American Federation of Teachers, Washington, D.C.

Daniel Goroff, Center for Science, Mathematics, and Engineering Education, National Research Council

Mary Gromko, State Science Consultant, Colorado Department of Education, Denver

Kathleen Guidroz, Commission on Behavioral and Social Sciences and Education, National Research Council

Norman Hackerman, The Robert A. Welch Foundation, Houston, Texas

Janet Hansen, Commission on Behavioral and Social Sciences and Education, National Research Council

Ruth E.S. Hayhoe, Ontario Institute for Studies in Education, University of Toronto, Canada

Cadelle Hemphill, Education Statistics Services Institute, Washington, D.C.

Deborah Inman, Policy Institute, Office of Education, Research and Information, U.S. Department of Education

Kathleen Johnston, Center for Science, Mathematics, and Engineering Education, National Research Council

Jeane M. Joyner, North Carolina State Department of Education, Raleigh

Eugina Kemble, American Federation of Teachers, Washington, D.C.

Michael Kestner, Mathematics and Science, North Carolina Department of Public Instruction, Raleigh

Jeremy Kilpatrick, Department of Mathematics Education, University of Georgia, Athens

Daniel Koretz, The RAND Corporation, Washington, D.C.

Gerald Kulm, American Association for the Advancement of Science, Washington, D.C.

Carole Lacampagne, Office of Education, Research and Information, U.S. Department of Education

Paul G. LeMahieu, Delaware Education Research and Development Center, University of Delaware, Newark

William Linder-Scholer, SciMathMN, St. Paul, Minnesota

Vena M. Long, Department of Curriculum and Instruction, School of Education, University of Missouri, Kansas City

Susan Loomis, American College Testing, Inc., Iowa City

Susan Loucks-Horsley, Center for Science, Mathematics, and Engineering Education, National Research Council

Megan Martin, Office of the President, University of California, Oakland

Michael Martin, TIMSS Study Center, Boston College

Michael Martinez, Department of Education, University of California, Berkley

Pamela Matthews, Department of Mathematics and Statistics, The American University

Marilyn E. Mays, Commission on the Future of the Standards, North Lake College, Irving, Texas

Patricia McAllister, Educational Testing Service, Washington, D.C.

Maggie McNeely, National Institute for Student Achievement, Curriculum and Assessment, Washington, D.C.

Patricia McWethy, National Science Education Leadership Association, Falls Church, Virginia

Karen Mitchell, Commission on Behavioral and Social Sciences and Education, National Research Council

Patricia Morison, Commission on Behavioral and Social Sciences and Education, National Research Council

Ina Mullis, TIMSS Study Center, Boston College

Mark D. Musick, Southern Regional Education Board, Atlanta, Georgia

Erika Otto, Committee on Economic and Educational Opportunities, U.S. House of Representatives, Washington, D.C.

Eugene Owen, National Center for Education Statistics, U.S. Department of Education

Christine Paulsen, Ph.D. Candidate, Graduate School of Education, University of Pennsylvania

Lois K. Peak, National Center for Education Statistics, U.S. Department of Education

John Pettit, National Security Agency, Ft. Meade, Maryland

Mattye Pollard-Cole, Colorado Department of Education, Denver

Andrew C. Porter, Wisconsin Center for Education Research, University of Wisconsin, Madison

Harold Pratt, Center for Science, Mathematics, and Engineering Education, National Research Council

Jack Price, National Council of Teachers of Mathematics, Newport Beach, California

Senta Raizen, National Center for Improving Science Education, The NETWORK, Inc., Washington, D.C.

Sarah Reinstein, Office of Senator Kennedy, U.S. Senate, Washington, D.C.

Barbara Reys, Department of Curriculum and Instruction, University of Missouri, Columbia

Heather E. Roberts, American Psychological Association, Washington, D.C.

Ron Rosier, Conference Board in the Mathematical Sciences, Washington, D.C.

Patricia O'Connell Ross, Office of Education, Research and Information, U.S. Department of Education

F. James Rutherford, American Association for the Advancement of Science, Washington, D.C.

Laura Salganik, Education Statistics Services Institute, American Institutes of Research, Washington, D.C.

Gerhard L. Salinger, Education and Human Resources Directorate, Division of Elementary, Secondary and Informal Education, U.S. National Science Foundation, Arlington, Virginia

Richard Schoen, Commission on the Future of the Standards, Department of Mathematics, Stanford University

Judith Segal, Office of Education, Research and Information, U.S. Department of Education

Yoshinori Shimizu, School of Education, Indiana University, Bloomington

Edward Silver, Learning Research and Development Center, University of Pittsburgh

Len Simutis, Eisenhower National Clearinghouse, Ohio State University, Columbus

Claibourne Smith, Advisory Council on Education Statistics, Technology and Professional Development Central Research Development, Dupont Corporation, Wilmington, Delaware

William Spooner, North Carolina Department of Public Instruction, Raleigh

Diane M. Spresser, Education and Human Resources Directorate, Division of Elementary, Secondary and Informal Education, U.S. National Science Foundation, Arlington, Virginia

John Staver, Center for Science Education, Kansas State University, Mahattan

Sharon Stenglein, Minnesota State Department of Education, St. Paul

David Stevenson, Office of the Undersecretary, U.S. Department of Education

John Supovitz, Horizon Research, Inc., Chapel Hill, North Carolina

Larry E. Suter, Division of Research, Evaluation and Communication, Education and Human Resources Directorate, U.S. National Science Foundation, Arlington, Virginia

John Sutton, Eisenhower High Plains Consortium for Mathematics and Science at McREL, Aurora, Colorado

Judith Sydner-Gordon, Los Angeles County Office of Education

Dan Taylor, National Assessment Governing Board, Washington, D.C.

Paul Thiess, Editor-in-Chief, *The Torch,* Washington, D.C.

Paul R. Trafton, Department of Mathematics, University of Northern Iowa, Cedar Falls

Roy Truby, National Assessment Governing Board, Washington, D.C.

Christopher VanWyk, Drew University

Gretchen Vogel, *Science*, Washington, D.C.

Lisa Weil, Office of the Governor, Denver, Colorado

Gerry Wheeler, National Science Teachers Association, Arlington, Virginia

Trevor Williams, Westat, Rockville, Maryland

James Wilson, Committee on Education, U.S. House of Representatives, Washington, D.C.

Walter Wingo, *Design News*, Arlington, Virginia

Janie L. Zimmer, Howard County Public School System, Ellicott City, Maryland

Appendix B

Symposium Agenda

LEARNING FROM TIMSS:
AN NRC SYMPOSIUM ON THE RESULTS OF THE
THIRD INTERNATIONAL MATHEMATICS AND SCIENCE STUDY

February 3-4, 1997

Program

Monday, February 3

Understanding TIMSS and Its Results

8:30 - 9:15 a.m.	**Welcome and Introductions** *Richard Shavelson*, Symposium Moderator, Stanford University *Bruce Alberts*, President, National Academy of Sciences and Chairman, National Research Council *Pascal Forgione*, Commissioner, National Center for Education Statistics *Luther Williams*, Assistant Director, National Science Foundation
9:15 - 10:30 a.m.	**The TIMSS Achievement Study** Moderator: *Robert Linn*, University of Colorado Presenter: *Albert Beaton*, Study Director, TIMSS International Study Center, Boston College
10:30 - 10:45 a.m.	*Break, anteroom*
10:45 - 12:00 a.m.	**The TIMSS Curriculum Study** Moderator: *Shirley Malcom*, American Association for the Advancement of Science Presenter: *William Schmidt*, Michigan State University, TIMSS U.S. National Research Coordinator and Director, U.S. National Research Center
12:00 - 12:45 p.m.	*Lunch, anteroom*
12:45 - 2:00 p.m.	**Explaining U.S. Performance** Moderator: *Richard Shavelson*, Stanford University Presenter: *Edward Haertel*, Stanford University Discussant: *Frederick Mosteller*, Harvard University

2:00 - 3:00 p.m.	**The Three-Country Videotape Study** Moderator: *Glenda Lappan*, Michigan State University Presenter: *James Stigler*, University of California - Los Angeles, Director of the Three-Country Videotape Study
3:00 - 3:15 p.m.	*Break, anteroom*
3:15 - 4:15 p.m.	**The Three-Country Case Study** Moderator: *Lynn Paine*, Michigan State University Presenter: *Harold Stevenson*, University of Michigan, Director of the Three-Country Case Study
4:15 - 5:15 p.m.	**Using Qualitative Studies for International Comparisons** Moderator: *Francisco Ramirez*, Stanford University Presenter: *Michael Huberman*, University of Geneva Discussant: *Deborah Ball*, University of Michigan

Tuesday, February 4

Beyond the Horserace: Using TIMSS Results for Policy and Practice

8:30 - 9:30 a.m.	**Using TIMSS in a World of Educational Change** Moderator: *Gail Burrill*, University of Wisconsin Presenters: *J. Myron Atkin*, Stanford University *Paul Black*, King's College of London Discussant: *Stephen Heyneman*, World Bank
9:30 - 10:30 a.m.	**Implications of TIMSS for Policy and Practice** Moderator: *Michael Kirst*, Stanford University Presenter: *Richard Elmore*, Harvard University Discussants: *Marshall Smith*, U.S. Department of Education *Daryl Chubin*, National Science Foundation
10:30 - 10:45 a.m.	*Break, anteroom*
10:45 - 11:45 a.m.	**Implications of TIMSS for Teaching** Moderator: *JoAnn Vasquez*, President, National Science Teachers Association Presenter: *Jan de Lange*, Freudenthal Institute, The Netherlands Discussant: *Mary Lindquist*, Columbus College
11:45 - 12:30 p.m.	**Summary Session** *Richard Shavelson*, Symposium Moderator

Appendix C

Papers Presented at the Symposium

J. Myron Atkin and Paul Black
Using TIMSS in a World of Educational Change. King's College, London, 1997. (Revised paper published in *Phi Delta Kappan*; see References.)

Daryl E. Chubin
Comments on "Implications of TIMSS for Policy and Practice" by Richard Elmore. National Science Foundation, 1997.

Jan de Lange
Looking Through the TIMSS Mirror from a Teaching Angle. Freudenthal Institute, Utrecht University, The Netherlands, 1997.

Richard F. Elmore
Education Policy and Practice in the Aftermath of TIMSS. Department of Education, Harvard University, 1997.

Edward H. Haertel
Exploring and Explaining U.S. TIMSS Performance. Stanford University, 1997.

Stephen P. Heyneman
Using TIMSS in a World of Change: Comments at the National Academy of Sciences. The World Bank, Washington, D.C., 1997.

Michael Huberman
Qualities of Instruction and Varieties of Qualitative Research. University of Geneva, 1997.

Appendix D

TIMSS Reports and Resources

Reports: Achievement Results

Harmon, Maryellen, Teresa A. Smith, Michael O. Martin, Dana L. Kelly, Albert E. Beaton, Ina V.S. Mullis, Eugenio J. Gonzales, and Graham Orpwood

1997 *Performance Assessment in IEA's Third International Mathematics and Science Study (TIMSS).* International Association for the Evaluation of Educational Achievement. TIMSS International Study Center. Chestnut Hill, MA: Boston College.

International Association for the Evaluation of Educational Achievement

1996 *Mathematics Achievement in the Middle School Years: IEA's Third International Mathematics and Science Study (TIMSS).* Center for the Study of Testing, Evaluation and Educational Policy. Chestnut Hill, MA: Boston College. Available: http://wwwcsteep.bc.edu/timss [June 19, 1997].

1996 *Science Achievement in the Middle School Years: IEA's Third International Mathematics and Science Study (TIMSS).* Center for the Study of Testing, Evaluation and Educational Policy. Chestnut Hill, MA: Boston College. Available: http://wwwcsteep.bc.edu/timss [June 19, 1997].

National Center for Education Statistics

1996 *Pursuing Excellence: A Study of U.S. Eighth-Grade Mathematics and Science Teaching, Learning, Curriculum, and Achievement in International Context.* Washington, DC: U.S. Department of Health and Human Services. Available: http://www.ed.gov/NCES/timss [July 8, 1997].

A general overview of TIMSS and summary of eighth-grade results for the U.S.

1997 *Pursuing Excellence: A Study of U.S. Fourth-Grade Mathematics and Science Achievement in International Perspective.* Washington, DC: U.S. Department of Health and Human Services. Available: http://www.ed.gov/NCES/timss [July 8, 1997].

A general overview of TIMSS and summary of fourth-grade results for the U.S.

Martin, Michael O., Ina V.S. Mullis, Albert E. Beaton, Eugenio J. Gonzalez, Teresa A. Smith, and Dana L. Kelly

1997 *Science Achievement in the Primary School Years: IEA's Third International Mathematics and Science Study (TIMSS).* TIMSS International Study Center. Chestnut Hill, MA: Boston College.

Mullis, Ina V.S., Michael O. Martin, Albert E. Beaton, Eugenio J. Gonzalez, Dana L. Kelly, and Teresa A. Smith

1997 *Mathematics Achievement in the Primary School Years: IEA's Third International Mathematics and Science Study (TIMSS).* TIMSS International Study Center. Chestnut Hill, MA: Boston College.

Reports: Curriculum

Howson, Geoffrey

1995 *Mathematics Textbooks: A Comparative Study of Grade 8 Texts.* TIMSS Monograph No. 3. Vancouver, Canada: Pacific Educational Press.

An analysis of textbooks from eight countries, based on research from the Survey of Mathematics and Science Opportunities.

Schmidt, William H., Curtis C. McKnight, and Senta A. Raizen

1997 *A Splintered Vision: An Investigation of U.S. Science and Mathematics Education.* U.S. National Research Center for the Third International Mathematics and Science Study. Dordrecht, The Netherlands; Boston, MA; London, England: Kluwer Academic Publishers.

A report of the curriculum analysis results for the United States, drawing on results of the international curriculum analysis and on data from the background questionnaires not otherwise available as of fall 1997.

Schmidt, William H., Curtis C. McKnight, Gilbert A. Valverde, Richard T. Houang and David E. Wiley

1997 *Many Visions, Many Aims: Volume 1: A Cross-National Investigation of Curricular Intentions in School Mathematics.* Dordrecht, The Netherlands: Kluwer Academic Publishers.

A report of the curriculum analysis results.

1997 *Many Visions, Many Aims: Volume 2: A Cross-National Exploration of Curricular Intentions in School Science.* Dordrecht, The Netherlands: Kluwer Academic Publishers.

Survey of Mathematics and Science Opportunities
1996 *Characterizing Pedagogical Flow: An Investigation of Mathematics and Science Teaching in Six Countries.* Dordrecht, The Netherlands: Kluwer Academic Publishers.
The primary report of the survey of mathematics and science opportunities.

Reports: Methodology

Robitaille, David F., William H. Schmidt, Senta Raizen, Curtis McKnight, Edward Britton, and Cynthia Nicol
1993 *Curriculum Frameworks for Mathematics and Science.* TIMSS Monograph No. 1. Vancouver, Canada: Pacific Educational Press.

Robitaille, David F., and Robert A. Gordon, eds.
1996 *Research Questions and Study Design.* TIMSS Monograph No. 2. Vancouver, Canada: Pacific Educational Press.

Third International Mathematics and Science Study
1996 *Third International Mathematics and Science Study: Quality Assurance in Data Collection,* M. Martin and I. Mullis, eds. Center for the Study of Testing, Evaluation and Educational Policy, Chestnut Hill, MA: Boston College. Available: http://wwwcsteep.bc.edu/timss [June 30, 1997].
1996 *Third International Mathematics and Science Study: Technical Report, Volume 1: Design and Development.* Chestnut Hill, MA: Boston College. Available: http://wwwcsteep.bc.edu/TIMSS1/TIMSSPublications.html#International [July 8, 1997].

Actual Test Items

TIMSS Mathematics Items Released Set for Population 2 (seventh and eighth grades): All publicly released items used to assess seventh- and eighth-grade students in the TIMSS study.

TIMSS Science Items Released Set for Population 2 (seventh and eighth grades): All publicly released items used to assess seventh- and eighth-grade students in the TIMSS study.

TIMSS Mathematics Items Released Set for Population 1 (third and fourth grades): All publicly released items used to assess third- and fourth-grade students in the TIMSS study.

TIMSS Science Items Released Set for Population 1 (third and fourth grades): All publicly released items used to assess third- and fourth-grade students in the TIMSS study.

To order, contact: TIMSS International Study Center, (CSTEEP), Campion Hall Room 323, School of Education, Boston College, Chestnut Hill, MA 02167. (617) 552-4521. Also, can be downloaded from: http://wwwcsteep.bc.edu/TIMSS1/TIMSSPublications.html#International

Videotapes

International Association for the Evaluation of Educational Achievement

1997 Examples from the Eighth-Grade Mathematics Lessons in the U.S., Japan, and Germany. VHS video available from the National Center for Education Statistics, U.S. Department of Education, Washington, DC.

1997 Examples from the TIMSS Videotape Classroom Study: Eighth-Grade Mathematics in Germany, Japan, and the United States. CD ROM Video NCES 97-198. Available from the National Education Data Resource Center, c/o Pinkerton Computer Consultants, Inc., Alexandria, VA. (703) 845-3151.

Third International Mathematics and Science Study

1997 A Video Report, February 1997. A 13-minute summary of eighth-grade findings with commentary. Available from the Superintendent of Documents. U.S. Government Printing Office, Washington, DC 20402. (202) 512-1800.

Resources

Information about TIMSS, as well as copies of published reports, can be obtained from the sources listed below:

TIMSS International Study Center
CSTEEP, Campion Hall 323
Boston College
Chestnut Hill, MA 02167
617/552-4521
http://wwwcsteep.bc.edu/timss

U.S. National Research Center
Michigan State University
http://ustimss.msu.edu

Texts of the seven TIMSS newsletters, which provide descriptions of components of the study, abstracts of the curriculum study reports, and ordering information are available in this web site.

Reports available from Kluwer Academic Publishers Group
Order Department
P.O. Box 358
Accord Station
Hingham, MA 02018-0358
617/871-6600
services@wkap.nl

National Center for Education Statistics, TIMSS Project
555 New Jersey Ave., N.W., Suite #402A
Washington, DC 20208
Telephone 202/219-1333
TIMSS@ed.gov
http://www.ed.gov/NCES/timss

Full texts of TIMSS reports released by NCES are available on this web site. NCES has also produced a resource kit, "Attaining Excellence," designed for public education and for teachers, local decision makers, curriculum planners, and parents. Modules on different topics can be ordered separately. Superintendent of Documents, P.O. Box 371954, Pittsburgh, PA 15250-7954. (202) 512-1800 or orders@gpo.gov/su_docs.

Reports available from the National Library of Education
555 New Jersey Avenue, N.W.
Washington, DC 20208
Telephone 800/424-1616 or 202/219-1736